William George Ward

The Relation of Intellectual Power to Man's True Perfection

William George Ward

The Relation of Intellectual Power to Man's True Perfection

ISBN/EAN: 9783337367329

Printed in Europe, USA, Canada, Australia, Japan

Cover: Foto ©berggeist007 / pixelio.de

More available books at **www.hansebooks.com**

THE RELATION OF INTELLECTUAL POWER

TO MAN'S TRUE PERFECTION,

CRITICISM IN THE *RAMBLER* 1862.

ARD, D. PH.

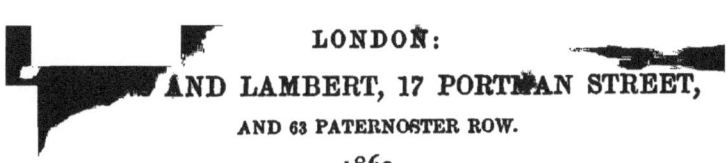

LONDON:
AND LAMBERT, 17 PORTMAN STREET,
AND 63 PATERNOSTER ROW.
1862.

PREFACE.

I TAKE for granted, that no one will take the trouble to read this pamphlet, unless he is desirous of thoroughly examining the controversy to which it refers. I will therefore here reiterate a request which I make in p. 3; I will entreat the reader to have my original pamphlet at hand, and refer to it at once whenever it is cited. I believe, of course, that the original work contained a statement of reasons, amply sufficient for its various conclusions; my chief business, therefore, in this, will be the drawing attention to those reasons. It would be simple waste of time and room, to put down again here, what has been fully stated there.

It is but reasonable, that the reader should also refer to my opponent's article whenever I quote it. And I should be extremely glad if, as soon as he has completed this pamphlet, he would again read through the adverse criticism as carefully as possible; to see if there be any argument of the least force, which I have omitted to consider. Indeed, I am very confident that the more carefully that article is read, in so much the lower estimation will its reasoning and conclusions be held.

As I am supposing that the reader really wishes to

judge on the whole case, I subjoin part of my original Preface; with no other verbal alteration, than that mentioned in p. 4 of the present pamphlet. He will thus have before him my own original statement, of the theses which I intended to advocate; and will be able to examine, step by step, how far my opponent has even attempted to argue against them.

"The first question which presents itself is this: Supposing God to have endowed any of us with" eminent philosophical power, "is it or is it not His wish, in so endowing us, that we should aim earnestly at making our" philosophical "labour an instrument for promoting our interior piety? is it or is it not true, that the more we do this, precisely so much the more do we fulfil the one end of our creation?

"The second question will be, whether I have truly stated, that one man makes a nearer approach than another towards fulfilling the end of his creation, in precise proportion as he is more intimately united to God by the pious dispositions of his will.

"The third question will be, whether I am correct in saying, that the answer, which I give to the two first, is implied for certain in the Church's doctrine and practice.

"The fourth question will be, whether I have rightly alleged that certain most serious dangers are to be apprehended, in the case of able and original thinkers, whose" philosophical labour "is not spiritually regulated and controlled.

"As these four questions are prior in scientific order, so also they are of far superior importance, to the remaining three. I may add also, that the first and fourth are of all the most fundamental and momentous.

"The fifth question will be, whether the doctrine, contained in my answer to the first two, be rightly and catholicly expressed, as I have expressed it: viz. by saying that man's perfection consists exclusively in the perfection of his moral and spiritual nature," philosophical "excellence having no part in it whatever."

"The sixth question will be, whether my use of the word 'intellect' is in accordance with the ordinary and popular acceptation of the term, here in England.

"The seventh and last question will be, whether those who appertain to what I have called the anti-Catholic schools of thought, really hold such opinions, as those attributed to them in my first Essay."

Finally, I would draw special attention to the short ninth section, in which the contrast is drawn out between my opponent's theory and my own.

London,
Feast of St. Joseph's Patronage, May 11*th*, 1862.

CONTENTS.

		PAGE
I.	Introductory	1
II.	Vindication of my first thesis. "Supposing God to have endowed any of us with eminent philosophical power, it is His Wish, in so endowing us, that we should aim earnestly at making our philosophical labour an instrument for promoting our interior piety. The more we do this, precisely so much the more do we fulfil the end of our creation"	4
III.	Vindication of my fourth thesis. "Certain most serious dangers are to be apprehended, in the case of able and original thinkers, whose philosophical labour is not spiritually regulated and controlled"	6
IV.	Vindication of my second thesis. "One man makes a nearer approach than another towards fulfilling the end of his creation, in precise proportion as he is more intimately united with God by the pious dispositions of his will"	7
V.	On the theological word 'intellectus'	8
VI.	Vindication of my third thesis. "My first two theses are implied for certain in the Church's doctrine and practice"	15
VII.	Theological unsoundness of the opinion, that, merits being equal, the keener human intellectus will see God more perfectly. Its irrelevance, even were it not unsound .	20
VIII.	Vindication of my fifth thesis. "The doctrine contained in my first two theses is rightly and catholicly expressed as I have expressed it; viz. by saying that	

man's perfection consists exclusively in the perfection of his moral and spiritual nature; philosophical excellence having no part in it whatever" . . . 30
IX. My opponent's theory contrasted with mine . 41
X. Conclusion 44

REPLY TO CRITICISM

IN

THE "RAMBLER."

I.

I RECENTLY published a brief inquiry, on "the relation between intellectual power and man's true perfection." This inquiry, as I think, does not derive its importance at all from the possibility, that good Catholics can differ as to its true resolution, when once the point at issue is clearly understood. Its importance arises from the fact, that the point at issue is *not* clearly understood; and that many thinkers, in their estimate of men and things, proceed implicitly on principles, of which, if explicitly stated, they would at once see the monstrous unreasonableness. These therefore are the two truths, which I am desirous of pressing strongly on the attention of Catholics in general. (1) I wish them to apprehend (and then of course they will accept) that one doctrine, on the relation between intellectual and spiritual excellence, which (if it is but clearly stated) no good Catholic can possibly doubt; and which indeed no Theist can doubt, without most manifest unreason. (2) I wish them also to observe, how constantly that one doctrine is forgotten and its contradictory implied, in various views and reasonings that are every day put forth.

A criticism on my pamphlet has appeared in the current *Rambler*, of which, I think, the most remarkable characteristic, is the cloud of confusion which it has thrown around the whole subject. Now if my principal theses had been therein clearly and intelligibly stated, I am confident that the instinct of all good Catholics would at once have recognised their truth; and I should have cared little therefore to answer adverse *arguments*. But the *Rambler* criticism would lead all, who have not seen my pamphlet, into the most complete misapprehension of its whole drift and bearing. I felt at once therefore, that a reply was imperatively necessary; though I hope it may not extend to any considerable length.

There is hardly any discussion, of equal importance with this, which may more easily degenerate into a sterile logomachy. And there was the greater danger of this result, because (as I stated in my Preface), through circumstances altogether unavoidable, "the arrangement of my two Essays is not only not scientific, but essentially the reverse." That I might avert this danger to the best of my power, I put forth a methodical enumeration, in that Preface, of "the various questions which present themselves for consideration." It was of course abundantly competent for my opponent to maintain, that this enumeration is no fair analysis of the Essays themselves. Yet even in that case, he was surely bound to accept it, as my own final and deliberate statement, of the questions which my little volume was intended to raise: whereas it is truly remarkable, that he has abstained from all allusion to that Preface, as absolutely as if it had never been written. He has made no attempt whatever, to grapple with my case as I myself presented it; but has merely ignored the fact of my having thus presented it at all. And by this means he has imported into the controversy (I am quite sure without any intention of his own) an amount of fallacy and misconception, which would have been simply impossible, had my own statement been accepted as the clue of my own argument.

I still retain the opinion, that my Preface presents my theses in the shape most available, for clearly apprehending the point at issue. I will therefore take, one by one, the various questions there recited; though I will somewhat vary their order, to meet the exigencies of the present controversy. And I have but two preliminary remarks to make, before entering on this task. (1) I will make the earnest request, that any reader who cares to examine the present argument, will keep constantly at hand my original "two Essays;" and that he will refer immediately to any passage which I may cite from them. A great deal of additional clearness will thus be gained, and a great deal of needless trouble avoided. (2) My second preliminary remark is this. I maintained in my two Essays, that there is a most wide and important difference of sense, between the word "intellect" as ordinarily used here in England, and the word "intellectus" as found in theological writers. My opponent denies this distinction; and (as I should say) his argument is one continued fallacy, founded upon that denial. I shall have to discuss the matter with him presently; but in the mean time, it will be necessary to keep clear of this constantly-recurring equivocation. I stated expressly, that I used the phrase "intellectual excellence" to express "philosophical or scientific excellence:" and my opponent admits, that this is its ordinary sense according to English usage.* Wherever therefore I quote from my former pamphlet, or cite my opponent's statements, I

---

* It will be seen presently, that my opponent's difference with me consists, in his maintaining that the theological word "intellectus" has a similar sense.

will change the word "intellectual" with its cognates, into the words "philosophical," "scientific," and the like.

In my Preface I laid down seven questions, as containing the principal points which my pamphlet was intended to raise; and, as I answer all of them in the affirmative, I will call those affirmative answers my seven theses. Of these, I stated that "the first and fourth are the most fundamental and momentous." With the first and fourth, therefore, I will commence.

II.

The first thesis is as follows: "Supposing God to have endowed any of us with" eminent philosophical power, it is "His Wish, in so endowing us, that we should aim earnestly at making our" philosophical "labour an instrument for promoting our interior piety. The more we do this, precisely so much the more do we fulfil the one end of our creation" (Preface, p. vi.). For the purpose of both explaining and vindicating this thesis, I referred to a long passage pp. 39-46. I will extend this citation to p. 47; and I will beg the reader, before he proceeds with my ensuing remarks, to peruse carefully the reference. I will beg him to begin with the second paragraph of the fourth section (p. 39), and go on to the conclusion of the fifth (p. 47). I expressly drew out this passage, as a statement once for all of the doctrine for which I was contending (see first paragraph of section 4); and yet my opponent has taken no more notice of it, than if it did not exist. I may briefly sum up its argument thus, so far as regards my first thesis.

If we are gifted with a mind calculated to attain great philosophical power, this circumstance affords a strong presumption, that God wishes us carefully to exercise and cultivate such power. No one, I suppose, will say that this is *more* than "a strong presumption;" no one will allege the supposition to be actually impossible, that (notwithstanding such endowment) God may really call us to a different mode of life altogether. I have no wish, however, to insist on this: let me suppose it as certain, in regard to all those who possess such capabilities, that God wishes them to give those capabilities persevering exercise and development. I proceed, then, to ask, *in what degree* are we to do this? In any given case, it will be abundantly possible to practise mental discipline of this kind so constantly and unremittingly, as to clash undeniably and unmistakably with other duties. No one, I take for granted, will advocate this. What *is* the degree, then, in which we should practise scientific cultivation? My answer (and the answer of every good Catholic) is simply this. By carefully pondering on all the circumstances of the case,—our various obligations, our external circumstances, our inward character, and the like,—we are constantly to aim at discovering God's Preference in the matter. We are to exercise and cultivate our scientific power, in that degree and under those conditions, which (so far as we can discover) will be most pleasing to our Holy Creator. But, in proportion as we do this, we are making our philosophical labour an instrument for promoting our interior piety: we act more rightly, therefore, in proportion as we make our philosophical labour an instrument for promoting our interior piety. But, in proportion as we act more rightly, we please the more our Holy Creator; or, in other words, fulfil the one end for which He created us. The inference is obvious. Here is my first thesis, and here are its grounds.

What has my opponent to object? I protest myself unable to discover. He says, indeed, that, on this view, philosophical excellence "is no part of the man at all; but is merely an instrument, superior to the muscles only as one tool may excel another" (p. 466). This is a strong expression of dislike to my thesis; but it contains no argument against it.

Then he says (p. 472), that "the *subordination* of" the philosophical to the "spiritual" "is not in question." On the contrary, it is the one thing which *is* in question.

Lastly, he says (p. 488), "God, having given us" philosophical power "as a part of ourselves, wills that we should obtain" philosophical "as well as moral excellence." As regards those comparatively few persons to whom God *has* given philosophical power, who ever thought of doubting this? The question is,—in what degree, and under what conditions, are we to labour for the attainment of such excellence? And to this question, the only relevant one, I can really find no answer at all in the whole article.

Now my opponent professes to prove (p. 468), that my "main position" is "utterly untenable." Yet it was this first thesis which (as I expressly stated) I held to be "the most fundamental and momentous of all" which I put forth.* And against this, he has not even professed to adduce one single argument.

III.

My fourth thesis is, that "certain most serious dangers," recounted in my two Essays, "are to be apprehended, in the case of able and original thinkers, whose" philosophical labour "is not spiritually regulated and controlled." I refer to the remarks in my former pamphlet, from p. 26 to p. 29, and from p. 54 to p. 71.

---

* I said "the first and fourth;" but the fourth is simply a corollary from the first.

Every one who reads these remarks, will admit that the allegations contained in them, if true, are practically important, in a degree which it is difficult to exaggerate. Yet my opponent, from first to last, has not said one syllable in opposition to their truth. I repeat, he has not so much as attempted one word of reply, to those two theses, which I urged as being "of all the most fundamental and momentous."

IV.

Let me next state my second thesis. "One man makes a nearer approach than another towards fulfilling the end of his creation, in precise proportion as he is more intimately united with God by the pious dispositions of his will" (Preface, p. vi.). The distinction between this and the first thesis, is sufficiently obvious. In the first, I consider any given individual; and compare one imaginable course of conduct with another: but in the second, my comparison is between man and man. At the same time, its principle is substantially identical with that of the former; nor has my opponent attempted to argue against it. I will here therefore but briefly recapitulate the grounds on which it may be based. God has given us the knowledge of Himself and of His rightful claims over our conduct. He has done this, of course, in order that we may duly recognise those claims, and defer to them. We more nearly approach, therefore, towards fulfilling the end of our creation, in proportion as we consult His Preference, as to every detail of conduct; as to our whole use of that life, and of those faculties, which He has given us. But in proportion as we are ever consulting His Preference in these respects, in that proportion we are united to Him by the pious dispositions of our will. No good Catholic, then, will doubt my second thesis.

V.

A vindication of my third thesis will occupy a far larger portion of space: not because it is at all more doubtful than the preceding, but because my opponent has made much greater profession of arguing against it. Before entering on this vindication, it will be necessary to consider my opponent's statement, on the meaning of that very important theological word "intellectus."

He admits me to be perfectly correct, in my own definition of "intellectus;" he agrees with me, that "we are said by theologians to exercise our intellectus, so far as we contemplate, in any kind of way, real or apparent truth" (p. 469). But he adds, that there is, in fact, no other way of contemplating truth, *except* the philosophical. "Every act of knowledge, if real, is a certain measure of philosophy." We do not "know a thing at all until we know its genus and differentia" (p. 470). The matter at issue then between us is this. He maintains that the intellectus better performs all its various processes, in proportion as the mind has received philosophical and scientific cultivation. For myself, on the contrary, I maintain these three propositions. (1) There are many processes of the intellectus, towards which it receives invaluable help from scientific cultivation; but (2) there are many others, towards which it receives no help at all from such cultivation; and (3) there is a third considerable class of actus intellectuales, to whose due performance scientific cultivation is a great impediment.* And in joining issue on this question, I

---

* Surely all thoughtful readers will think that the following statement of Mr. Stuart Mill's is far nearer the truth than my opponent's. He refers to "the tendency of scientific pursuits in general; the influence of habits " of *analysis and abstraction* upon the character. How, without those habits, " the mind is the slave of its own accidental associations, the dupe of every

will begin by arguing it without any reference to God and the Invisible World.

In my second Essay (p. 37), I mentioned the poetical faculty, as an important portion of the intellectus. Let me take from it my first illustration. My instances of its exercise, indeed, may probably excite a smile in any poetical man who may see them; for I have (alas!) no poetical appreciation whatever: but the poorness of my instances will in no way affect the strength of my argument. Take, then, such judgments as the following: "This passage is most truly sublime;" "that description is too full of conceits, and is destitute of simplicity;" "this beautiful scenery is treated far more poetically by A B than by C D." These are judgments formed by a reader of poetry; others are formed by the poet himself: "This poetical idea is relevant to my theme, and will greatly enliven it;" "such words will express beautifully an idea, which cries aloud for beautiful expression." These, and a thousand other such judgments, are formed, whether by poets or by readers of poetry; and it is beyond question a very high excellence of the intellectus, to form them truly, and with promptitude of suggestion. Is any one so insane as to say, that scientific cultivation

---

"superficial appearance. . . . On the other hand, how their exclusive "cultivation, . . . by accustoming the mind to consider in objects, "chiefly the properties on account of which we refer them to classes and "give them general names, *leaves our conceptions of them as individuals lame* "*and meagre.* How, therefore, the corrective and antagonist principle . . . "is to be sought in those pursuits which deal with objects in the concrete; "clothed in properties and circumstances; real life in its most varied forms; "*poetry and art in all their branches.*"—*Dissertations and Discussions*, vol. i. p. 104.

I will but add, that, on the most important matters of thought, a good Catholic is preserved from being the "slave of accidental associations" and the "dupe of superficial appearances" by a far more potent instrumentality than "habits of analysis and abstraction."

will assist it in this task? that a "knowledge of genus and differentia" is the main thing needed? nay, will any one deny, that scientific habits are in themselves a serious impediment? though an impediment, of course, which may be in greater or less degree removed.

Let us pass on to the kindred judgments of *taste*: though here, again, my instances will be very meagre, for of taste I have less even than of poetry. "Cicero's choice of words is exquisite; and he locates them to the best advantage: but he is much too diffuse." "The imitations of classical Latin at the time of the Renaissance, are always readily distinguishable from genuine classical compositions." What help will my intellectus obtain from science or philosophy, in its important work of forming such judgments accurately and readily? What place is there here for "genus and differentia"?

In some degree analogous to taste, is perception of *character*. "I should not venture to propose this to A B; he would at once take offence: but I think we may speak of it to C D." "I could not trust E F with such a commission; but G H will be the very man." One can hardly imagine a droller spectacle, than a first-rate philosopher being called on to exercise such judgments as these. The very perfection, to which he has brought one limited province of his intellectus, brings down the same intellectus to the level of a child's, in other provinces no less important.

There is a knowledge too, not of character in general, but of its individual manifestations. A B is extremely quick in observing the *mood* of those with whom he converses; and sees at once when he is beginning to grow tiresome or to give offence. But C D drones on in his own strain; and is so immersed in theories or in self, that he is blind to all manifestations of weariness in his hearers. Ten to one, he is a man of rare philosophical attainment. He is so occupied with taking care that his words shall precisely express his ideas, and with giving an exact logical answer to his bored interlocutors' languid remarks, that he never thinks of observing their gestures, or of looking at their countenance. It would have been a great excellence of his intellectus, had it been ready in observing these facts, and drawing from them a true inference. But it has been so drilled in "genus and differentia," as to be quite incapacitated for this important service.

One more illustration will suffice. A common sailor has one extremely great virtus intellectualis: he is able to prognosticate a storm or a calm, and to draw other invaluable inferences, from phenomena which to a landsman have no significance at all. I should be very sorry to take a dangerous voyage in some ship, of which the various officials should be deeply interested in "genus and differentia." I fear they would be spinning theories, instead of watching phenomena; and their intellectus would lose this singular excellence, of prognosticating future danger.

I might fill any number of pages with further illustrations; but I have said quite enough for my purpose. I have shown unmistakably, that there are very many excellences of the intellectus, and those too of extreme importance to man's temporal well-being, which (so far from being promoted) are most seriously injured by philosophical cultivation. Such cultivation is one very great excellence of the intellectus; but it is only one out of many. But now as to that which is immeasurably the highest dos intellectualis, the power of contemplating supernatural truth. And here I will first notice my opponent's objection, that I "exclude from man's perfection" scientific " excellence in theology as well as in physics" (p. 469). I cannot understand his meaning. Scientific excellence is one and the same power, however exercised. It may be applied to the elucidation of revealed truth; or of moral truth in the natural order; or of metaphysical, physical, or historical truth. It may be directed to a good or a bad end; to the service of God or of the devil. But the habit of mind in itself is one and the same. The very same man, who, from natural premisses, will reason acutely and philosophise both largely and accurately, will do the same from supernatural premisses also; adequate knowledge and apprehension of those premisses being supposed. I believe no one will think of denying this, and no more therefore need be said.

I will next criticise my critic in another particular. "The intellect of the devil," he remarks with singular moderation of speech, "even as an intellect, *can never be perfect*, because its powers can never be exercised on the fullness of truth. The intellect* of the lost *can never be perfect*, for they can never know the chief good" (p. 471). The intellectus of evil spirits is certainly "not perfect," but rather in a state the most miserably calamitous; for it practically represents to them the Most Holy God as an odious oppressor, and every kind of foul enormity as a desirable end of action. But as to their 'intellect' in my opponent's sense of the word,—their power of analysing, reasoning, generalising, philosophising,—it cannot possibly be denied, that in these respects they very greatly excel mankind, because of the far greater intrinsic power of the angelic nature.† I repeat what I said in my former pamphlet. If scientific excel-

---

* Printed " will" by a droll mistake.
† See Suarez's chapter " de Angelis," lib. viii. c. 6, with the authorities there cited.

lence is in itself "intrinsically great and noble," then "a quality is in itself great and noble, which is possessed by the devils in a greater degree than by the ablest of us all" (p. 29, and Preface, p. xi.). I beg my opponent's careful attention to this argument; which as yet he has rather evaded, than fairly attempted to meet.

I now proceed with my remarks, on the mode whereby the intellectus is best enabled to contemplate supernatural truth. I am very far from denying, that philosophical and theological studies may be made most useful for our advancement in spirituality. "By means of theological study," I observe (p. 51), "we are able to obtain a surer and firmer grasp of supernatural truth; and we may derive great help from" philosophical discipline "in examining our motives of action more accurately; seeing our faults more clearly; devising more judicious means of spiritual improvement." Nay, I went further than this. I gave my opinion (p. 52) that "where equal promptitude of will" exists, the more educated man will be the more pious, because those *acts* of obedience and service to which he is prompt are more vigorous and sustained. My opponent, by the way, has ludicrously misapprehended this opinion of mine (p. 478); though it is not worth while to say more on the matter.

It cannot be said then, that I have ignored the advantage of scientific ability and cultivation, towards the attainment of religious knowledge. But if it be maintained, that such ability and cultivation is our *principal* means of apprehending Divine Truth, I should most indignantly protest. I should maintain with the most undoubting confidence, that interior piety confers an immeasurably keener and truer apprehension of the great Verities, than is obtainable by the mere help of science. Or in other words I should maintain, that an uneducated

Saint will have an immeasurably deeper and more vivid appreciation of them, than the most accomplished theologian who is not saintly.

On the extremely opposite ground to this, stands my opponent. I have no wish at all to take advantage of an incautious expression; but there is one sentence in his article, to which I would earnestly draw his attention. It occurs in p. 472. I had said that "in proportion as we grow in perfection of will, we grow in perfection of intellectus; for we apprehend supernatural truth more keenly and vividly." He replies, "I do not see how any quality of the will as such can render the intellect keener; unless" it be meant "that obedience merits grace and so enlightenment." This certainly will be understood to state, whether my opponent so intends it or no, that sensuality and worldliness do not, in the way of natural consequence, dim our spiritual sight; and that the life of faith and interior mortification has no intrinsic tendency, towards clearing and intensifying our contemplation of Divine Truth. This is far too serious a question to be treated episodically; nor can I bring myself to believe, that he really means what he seems to say. I will here therefore for the present only refer to the two writers, whom I before cited on the dona intellectualia: viz. St. Thomas and Lallemant.

As to St. Thomas, let me refer to my remarks in pp. 82-84 of my former pamphlet. These have been left unnoticed by my opponent, and I hope my readers will refer to them before proceeding further. St Thomas's doctrine then is this. There are two kinds of wisdom; wisdom the donum, and wisdom the virtus intellectualis: and the former is a higher gift than the latter. The latter arrives at a right judgment on things divine, by the inquiry of reason; but the former through that sympathy with the supernatural, which is caused by the virtue of charity. According to St. Thomas therefore, the virtue of charity is a higher and more important way, for obtaining rectitude of judgment on things divine, than is scientific inquiry and investigation. I challenge my opponent to face this doctrine of St. Thomas's. As to Lallemant, he shall speak for himself.

"A soul which, by mortification, is thoroughly cured of its passions, and by purity of heart is established in a state of perfect health, *is admitted to a wonderful knowledge of God* and discovers things so great," &c.—p. 152.

"By purity of heart we enter gradually into the various meanings which [Scripture] contains: and although we may have read it through a hundred times, yet *if we still make progress in purity of heart* and persevere in the study of it, we shall *penetrate its mysteries continually more and more.*"—p. 159.

"If difficulties of conscience are proposed to" those who have the donum scientiæ in its largest share, "they will give an admirable solution. Ask them for the reason of their reply, they cannot tell you, because they know it without reasoning by a light superior to all reason."—p. 164.

"A pure soul will learn more in one month by the infusion of grace, than others in several years by the labour of study."—p. 169.

I repeat therefore my former conclusion. The intellectus is then most perfectly constituted, when it is fittest for the contemplation of supernatural truth. But our intellectus is more fitted for such contemplation, mainly in proportion as we purify and mortify our will, and fix our thoughts on the unseen world. Perfection of intellectus therefore and perfection of will proceed pari passu together.

VI.

I have now cleared the way, for a discussion of my third thesis. But before entering on that discussion, let me write down the two first, which I consider myself to have already established.

(1) Supposing God to have endowed any of us with eminent philosophical power, it is His Wish, in so endowing us, that we should aim earnestly at making our philosophical labour an instrument for promoting our interior piety. The more we do this, precisely so much the more do we fulfil the one end of our creation.

(2) One man makes a nearer approach than another towards fulfilling the end of his creation, in precise proportion as he is more intimately united to God by the pious dispositions of his will.

The "third question," which we are now to consider, is, "whether I am correct in saying that" these two theses "are implied for certain in the Church's whole doctrine and practice." And I will first recapitulate the reasons given in my former pamphlet, for answering this question in the affirmative.

(1) Let me refer again to the long passage from pp. 39 to 47. This passage, I consider, shows most abundantly, that the two theses not merely are true, but are implied for certain in the ascetical principles received throughout the Church. They are in fact but the application of those general principles to one particular case.

(2) I have referred to St. Ignatius's Foundation. And I will beg the reader, before proceeding further, to look at my former pamphlet from p. 7 to p. 9. This Foundation then is (as I may say) infallibly declaratory of the Church's mind: and according to its teaching, we fulfil the end of our creation, in precise proportion as we "praise the Lord our God, and show Him reverence, and serve Him, and by means of this save our soul." Further, according to this Foundation, we act more rightly, the more *in all other things* we "desire and choose those which are more expedient to us for our true end." Now philosophical cultivation is unquestionably

one of those "all other things." Hence the Church teaches, that we act more rightly, in proportion as we desire and choose philosophical cultivation, precisely to that extent and under those conditions, which are more expedient to our true end: that end being, the praise, reverence, and service of God, in order to our soul's salvation. See also the quotation from Suarez, in the note to p. 50. This argument, which I put at the very outset of my case, is left by my opponent practically unanswered; as any one may see, who will read his remarks in p. 488.

(3) My next argument was from the consent of St. Thomas and all theologians. On this head, see also the quotations in the note to p. 48, and the remarks from p. 80 to p. 82. According to all theologians, "it is charity," not philosophical ability, "which unites us to God Who is our true End." So long as we live on this earth, we more fulfil our true end (according to the judgment of all theologians) in precise proportion as we have more charity. This statement is fully admitted by my opponent (p. 489): though how he reconciles such admission with his own doctrine, passes my comprehension.

(4) Next I argued from the Catholic doctrine on merit. I will beg my reader here to peruse the passage in pp. 10, 11. My opponent replies (p. 489) that "God only approves *for reward* free acts of the will; *for such alone are in our power to give or withhold*: but He approves *all that He sees in us of good*;" including, of course, philosophical power. My opponent here (strange as it must seem) actually speaks, as though those countless and protracted actus intellectuales, which generate philosophical power, were necessary and not free acts. I need hardly say, that Catholic Theology, speaking in accordance with most manifest reason, declares the exact contradictory. Every act, consummated in the intel-

lectus, is perfectly free; because the will has full power to direct the intellectus into this or that occupation. Every one, I say, of those innumerable actus intellectuales, which generated my philosophical power, had a morality of its own: it was morally good; or morally bad; more or less good; more or less bad; according to its intrinsic character. Its morality depended (1) on the circumstances under which it was elicited; (2) on the end at which the will aimed, in issuing its command to the intellectus; (3) on the greater or less efficacity with which the will adhered to that end. Supposing all these actus intellectuales had been morally bad, so far from God approving that philosophical power which is their result, He would simply abhor it. Take the ordinary case, that some were good and some were bad, He both approves and disapproves it: He approves it, so far as it was produced by acts morally good; disapproves it, so far as it was produced by acts morally bad; and would have approved it still more, had it been produced by acts morally better. This fourth argument of mine, therefore, does but derive greater strength, from my opponent's attempt to answer it.

(5) My remaining argument rested on the canonisation of Saints (see p. 11). Those Saints in Heaven, whom the Church proposes for our veneration, are undoubtedly those men who on earth have best fulfilled the end of their creation. And what has been the characteristic of them one and all? That they have devoted their lives, with heroic energy and perseverance, to the service of God; and that they have made all their earthly avocations (scientific or otherwise) so many instruments, for growing in purity of heart and love of their Creator. The conclusion is obvious; nor has my opponent one word to say against it (see p. 490).

As I said in my former pamphlet, I could multiply proofs of this third thesis quite indefinitely; but any one of the five just cited, though it stood alone, would be a complete demonstration. And now to consider what my opponent has to say in reply. Very little indeed, that is worthy of attention. He instances the doctrine of the Incarnation; of Mary; and of Adam. Now these doctrines, I admit, have a certain superficial appearance of inconsistency, with my mode of *expression* as to human perfection: though in my eighth section I shall easily show, that even in this respect they have no value whatever. But as against this third thesis of mine which I am now defending, they have not the slightest degree even of apparent plausibility. I am now virtually maintaining this thesis: " It is implied for certain in the Church's whole doctrine and practice, that one man approaches more nearly than another towards fulfilling the end of his creation, in precise proportion as he more loves God; in precise proportion as he makes all his various employments (scientific or otherwise) so many instruments for promoting his interior piety." It is not possible so much as to state in words any objection to this thesis, which shall be grounded on any of the above doctrines. Consider for instance our Blessed Lady. So far as she was occupied at all with scientific exercises, no one doubts that she used them as instruments for promoting her interior piety.

There is in fact but one objection raised by my opponent, so far as this third thesis is concerned, which even admits of being put into words. As he lays much stress on this objection, and as (if treated at all) it must be considered at some length, I will devote to it a separate section.

VII.

Certain theologians have thought, that of two men dying with equal merits, he who possesses the keener intellectus will see God the more perfectly. And it might seem to follow from this opinion, that the mere cultivation of philosophical power will in itself obtain for us a higher degree of beatitude; that it will obtain this, quite apart from the question, whether such cultivation has been subordinated to spiritual regulation and control. I confidently make three answers. (1) The opinion is one which no Catholic is at liberty to hold. (2) If it *were* a permissible opinion, it cannot by possibility lead to any conclusion, at variance with the doctrine for which I contend. (3) Though it *were* permissible, and though it *did* lead to consequences at variance with said doctrine, it would be monstrously unreasonable on that account to cite its advocates as opposed to me. And I will call the opinion (merely for clearness' sake and by way of distinction) Mastrius's opinion; for he holds it more explicitly and consistently, than any other theologian whom I happen to know.

First then I maintain, that no Catholic may lawfully hold Mastrius's opinion; for the Council of Florence expressly decrees, that those men who gain Heaven " intueri Deum sicuti est, pro meritorum tamen diversitate alium alio perfectiùs." I am very well aware that there may be much theological argument, on the precise force of the word 'merits' in this decree; but no one would ever dream of including, under the term, the nativa vis intellectûs. This vis intellectûs is, I suppose, as different in different men, as the degree of merit which they respectively possess at death. If Mastrianism therefore were true, nothing would be more common in Heaven, than that beatus A, greatly inferior in merit to beatus B, enjoys nevertheless greater glory, because of his far keener intellectus. Now there are some matters of fact, which at last are too plain to admit of argument; and surely here is one of them. If it happens, in a vast number of cases, that he who has less merit sees God more perfectly than he who has more, it cannot possibly be true, that "one man sees God more perfectly than another *according to the diversity of merits*." I repeat what I said in my former pamphlet (p. 93), that an Arian might on such terms subscribe the Nicene Creed, and a Lutheran accept the decrees of Trent. And all the Mastrians, whom I know, seem possessed by this consciousness; for not one of them has ventured to face that undeniable conclusion, to which I have now drawn attention. Not one of them, I say, has ventured expressly to state, that A's merits may be much inferior to B's, and yet his glory greater.

I shall have much to say presently, in further corroboration of my argument; but first let me consider what my opponent advances in objection to it. (1) He begins (p. 479) by declaring it to be a condemned proposition (viz. Baius's 14th) that "absolute position in Heaven depends solely on merit." This was probably said in haste: at all events it is a complete mistake. No condemned proposition can possibly be understood, except by examining its connexion with the condemned system; and in Baius's case the Pope expressly refers us to the "sensus ab auctore intentus." Now nothing is more certain, than the "sensus ab auctore intentus" of this proposition. Indeed if my opponent will but read the proposition which precedes this, and that which follows it, he will admit his misconception. Baius, in accordance with his whole theory, considers that "the good works of just men" "by the just judgment of God deserve to receive no larger reward" than that due to them as mere obedience, apart from their supernatural character: and in this 14th proposition he declares, that they will *in fact* receive no larger reward, than is due to them in that non-supernatural respect.*

(2) Secondly, my opponent says: "I suppose" that the place of the Holy Innocents in Heaven "as martyrs is higher than that of thousands who have accumulated vast stores of merit." This proposition, if true, would be so plausible an argument in his favour, that it should have been warranted by some authority stronger than "I suppose." I am very far from opposing my own ignorance to another man's knowledge; I only wish to know the grounds of that knowledge. Negatively, I can but say that I never before heard of such a supposition; and positively, that I have now consulted the index of Benedict XIV. 'de Canonisatione,' and can find no hint of it. He refers to the Holy Innocents, and to other infant martyrs (l. i. c. 14, n. 3; l. iii. c. 15, n. 4). He mentions an opinion, opposed by St. Thomas, that God endowed the Holy Innocents with an anticipated use of reason, in order that they might be martyrs in will as well as in act. He recounts various privileges, enjoyed by infant martyrs. But of the doctrine mentioned by my opponent I can find no trace.

(3) My opponent next quotes Luke xix. 24. I reply firstly, that we may not interpret Scripture, except in

---

* Suarez points this out exactly as I have stated it. De Gratiâ, l. xii. c. 31, n. 1. A little further on, he makes another remark, which I may as well quote: he is alluding apparently to Bellarmine, but tacito nomine. "Etiam si quis asserere velit, dari sanctis in patriâ aliquod gloriæ aug- "mentum ultra id totum quod meritis debetur, in primis non de qui- "busdam tantùm sed de omnibus id asserere oportere .... deinde illum "excessum .... esse inæqualem simpliciter *cum quâdam æqualitate pro-* "*portionis.*"—n. 7.

accordance with the Church's defined doctrines; secondly, that in the case of a parable (as all commentators agree) we can never know that this or that particular detail has any doctrinal meaning at all; and thirdly, that the parable cannot possibly bear the sense which my opponent supposes, and that Suarez says he can find no one who gives it that sense. I think Suarez's exposition altogether satisfactory; and refer therefore to his work De Gratiâ, l. xii. c. 21, n. 8.

(4) My opponent further maintains (p. 480) that Mastrius's opinion is in fact the general "Scotist doctrine on the subject." And he deservedly lays much stress on this allegation; for if it *were* a fact, it would be a very important one. I wish I could profess any large acquaintance with Scotist writers; but I can only say, that Mastrius himself tells a very different story. He says expressly that some Scotists are with him, and some against him; nor does he venture to insinuate, that the majority are on his side.* This latter fact alone will speak most clearly to those, who know the constant proclivity of scholastic theologians, to regard their own opinion as the more common. But a still more significant phrase occurs in his disputation. He maintains, as becomes a good Scotist, that his master Scotus is not really against him; and commemorates Smisingius's prowess, in vindicating from the enemy one of Scotus's passages. "Quare falsum est," he adds, "quòd ibi Doctor loquitur de habentibus meritum æquale; *ut communiter intelligunt alii Scotistæ*," n. 214. "The other Scotists," I infer, are "commonly" against him.

---

* [Controversia] est *inter Scotistas:* quidam enim, licet concedant de possibili, et de eo *quod eveniret ex naturâ rei*, posse intellectum perfectiorem cum æquali lumine elicere perfectiorem visionem, id tamen *negant de facto;* alii verò nedum de possibili, sed etiam de facto, contendunt ità rem se habere."—De Visione Beatâ, n. 193.

(5) My opponent further maintains, that the Mastrians of other schools are not so few as I suppose; and that I am mistaken in thinking they refer principally to the comparison between men and angels. He has said nothing to change my opinion on these two heads; but as neither of the points is essential, for brevity's sake I will say 'transeat.' In regard to the second point, I will but draw his attention to my quotations from Henno and Frassen at p. 89; which he has quite omitted to consider.

(6) But his main argument turns on the fact, that Mastrius's opinion, though maintained in the schools, has not been censured. See p. 480. I replied by anticipation to this reasoning in pp. 90, 91; and as my opponent has not even noticed my replies, I might well content myself, with merely begging the reader carefully to peruse that passage. After he has done so however, I will beg him to consider one or two further arguments.

And my first additional argument shall be, that my opponent's principle is quite unheard of in Theology. From the time Mastrius's opinion began to be promulgated, there has been a constant opposition to it in the schools, resting partly on other grounds, and partly on the decree of Florence. Did any later Mastrian ever argue, that his opinion was certainly not contrary to the Church's doctrine, because it had been so long uncensured? Yet if my opponent's principle had been ever heard of, this would have been the very strongest answer to such objections, that could possibly have been made.

But now for an argumentum ad hominem. It is absolutely certain, that many theologians have opposed Mastrius's opinion, as theologically unsound, and as opposed to the Council of Florence. It is absolutely certain also, that not one of them has been censured for doing this. On my opponent's own principle therefore,

"I have full right of holding" what they hold; viz. that the opinion which he advocates is directly contradictory to an infallible decree of the Church. Nor can he complain therefore, if I exercise that right.

But finally it may be asked, if the Church have really defined this truth, is it probable it can have dropped out of her practical teaching? This brings me round again to the point from which I diverged; viz. the positive evidence which exists of the Church's doctrine, that the Beatific Vision will be more or less perfect, exactly as each man's merits are greater or less. I said in my former pamphlet (p. 88), "I doubt if there is any one catechism, expounding doctrine in any degree of detail, which does not state this truth as a matter of course; as a most certain and unquestioned part of Catholic doctrine." My opponent replies (p. 479) that this "appeal . . . scarcely merits notice." If my statement be true, it follows, that every bishop throughout the Church teaches his flock, as an unquestionable verity, that doctrine which my opponent rejects. And this, it appears, is an argument, which "scarcely merits notice"!

Now how far is my statement borne out by facts? I will give a few samples, of the language held in the most approved and authoritative catechisms. Of these, none certainly possesses higher authority, than Ferreri's truly admirable exposition of doctrine.

"Those," he says, "who shall have most merited on earth, will be recompensed by God with greater glory in Heaven: and this superiority (maggioranza) consists in the more perfect possession of the Vision of God. . . . . All will be recompensed *according to their capacity*: and *those who have most merited will have most capacity* (saranno più capaci), and so will have greater glory; he who shall have less merited will have less capacity, and so will have less glory."—Part i., lesson 35.

My next citation shall be from Raineri's course of

familiar instructions, given in Milan Cathedral. I translate from the French translation:

"All, it is true, will not enjoy the same degree of glory; *that glory will be proportioned to their merits.*"—Part i. lesson 29.

From Italy I proceed to France. I have been told that Guillois's is the best French catechism: I have therefore referred to it.

Q. "Is the beatitude of the Saints the same in all?" A. "No: it is greater or less, *accordingly as on earth they have acquired greater or less of merits.* . . . . Does not distributive justice demand, that those who have been constantly faithful to the Lord, and who to prove their love to Him have imposed on themselves the greatest sacrifices, should receive a more abundant recompense, and be raised to a higher degree of glory, than those who have loved him but feebly? . . . . Accordingly St. Paul teaches that 'each man shall receive his particular reward according to his labour;' and has not the Saviour Himself said, 'in My Father's House there are many mansions'? That is to say, that there are some more resplendent (brillantes) than others; and *every thing there is proportioned to the merit of those who are called to occupy them.*"—Part i. lesson 30.

The Catechism of Montpelier teaches the same truth.

Q. "Are there not different degrees of glory?" A. "Yes: those who on earth shall have *more loved God and more perfectly imitated Jesus Christ*, shall be in a higher degree of glory."—Part i. sec. 2, c. 3, n. 20.

I will now cross the Channel, and refer to a work, with which my opponent may not improbably be well acquainted. It is a most excellent volume, and likely to be of far more important service, than many with much greater pretensions. It is called a *Manual of Instructions on Christian Doctrine*, and was published last year. On the doctrine before us, its statements are no less clear and unmistakable, than those of the Italian and French works which I have cited.

"Though all the Saints are supremely happy, there are various degrees in their bliss, *corresponding to the various degrees of merit which they have acquired in this life.* Not only *will those who have acquired greater merit be crowned with greater honour and glory*, but this glory will be different according to the special and peculiar virtues which they have practised."—p. 103.

I call upon my opponent to name, if he can, one single catechism, through the length and breadth of Catholic Christendom, which enters at all on the question of the different degrees of Beatific perfection, and which represents those different degrees as existing in any other proportion, than simply and precisely according to diversity of merits. If he cannot name one such catechism, he is plainly obliged to confess, that his doctrine is opposed to the universal teaching of the Catholic Church.

So much on her present teaching; next for her immemorial tradition. On this head, there cannot be a more unexceptionable authority than Petavius; and the following passage will speak for itself.

"Fulgentius in libro de Trinitate '*tantùm ampliùs*' ait '*alio* in illâ vitâ Deum videbit, *quantùm ampliùs in hâc vitâ Eum dilexit præceptisque Ejus obediens fuit.*' . . . . Hoc ipsum omnia illa Scripturæ ac Sanctorum testimonia persuadent, quæ unicuique justorum ac sanctorum, pro meritorum ac laborum modulo, consentaneam felicitatis æternæ mercedem esse persolvendam adstruunt : *quæ sunt innumerabilia.*"*—De Deo, l. vii. c. x. n. 4.

---

* It may be asked, how far the author alludes to any opinion like the Mastrian, as existing among the Fathers. As Petavius's works are so generally accessible, I will merely beg the reader to follow this very chapter to the end; and I will here give a brief account of what he will find. Petavius says that the scholastics have debated at great length, whether the "naturalis intelligentiæ vis ac mentis acies" confer something to the perfection of that Vision ; and whether, consequently, *angels see God better than men see Him.* But he adds, that extremely little is found about these controversies in the writings of the ancients. He quotes St. Cyril, however, and St. Gregory Nazianzen, as holding (the latter with some doubt) that the higher order of angels see God more perfectly than the lower. He also quotes a passage to the same effect, from the work on the " Celestial Hier-

Catholic tradition, then, is peremptory against Mastrius. And I may further add, that the very fact of comparatively few Scotists having held his opinion, is (if the case be so) an extremely strong fact in the controversy. No one can doubt, that his opinion is that which obviously results from the general Scotist doctrine; from the doctrine (I mean) that, wherever special intervention be not exercised, the keener intellectus under equal lumen will see God more perfectly. Great numbers (it would appear) of the Scotist school have exhausted every theological resource, in order to avoid drawing the Mastrian conclusion from the general Scotist premiss. What can have compelled them to this, except their deep sense of the repugnance which Mastrianism presents, to undoubted Catholic doctrine?

I protest, therefore, against being characterised by my opponent as a "dogmatising theologian" (p. 479).

---

archy," attributed to St. Dionysius. In addition to this, there is another passage from the same writer, which, without our having the context to guide us, is somewhat obscure. (1) It may refer only to angels; for the word *νοες*, used in it, is the proper word (Petavius tells us) for angels. (2) It may be spoken of men as well as of angels; and refer to that doctrine, just now cited from Ferreri, on man's 'capacity' for glory being proportioned to his merits. (3) It may imaginably mean, that one man will receive greater glory than another, in exact proportion to his mental vigour. According to either of the two former interpretations, its doctrine is not in the least Mastrian. According to the third, the writer would exclude merit from having even a share in determining our degree of beatitude: and my opponent will himself admit, that this statement is so undeniably anti-Catholic, as to deprive a passage, which should contain it, of all possible weight. We may be very sure, indeed, that no Catholic writer ever so expressed himself; and to me the *second* interpretation seems of all the most probable. St. Maximus's comment also runs far more intelligibly on that hypothesis.

This is the only one passage, cited by Petavius from the whole patristic period, of which it can be even doubtful, whether it is not in complete accordance with what I maintain to be the one Catholic doctrine. Surely there are very few truths, defined by the Church in later ages, on which such very harmonious and unmistakable agreement has existed throughout the earlier.

The doctrine that one man's future glory will be greater than another's accordingly as his merits are greater, was universally held in the early centuries; was defined at Florence; and has been ever since authoritatively taught throughout the Church. I am not giving a sense of my own to the Florentine decree, but interpreting it by universal tradition and by the Church's present teaching. I have already urged, in addition to all this, that its words absolutely defy my opponent's attempts at their interpretation. I infer, that the opinion, which he advocates, "is theologically unsound, even if it do not deserve a still severer censure."

Secondly, even if Mastrius's opinion were permissible, it leads to no consequences at variance with my thesis. No theologian has so much as hinted, that any *acquired* power of the intellectus, whether philosophical or other, will in itself augment our beatitude: the whole question regards what Petavius calls the "*naturalis* intelligentiæ vis"; the natural power, possessed by the intellectus, of contemplating God. At least, if there *be* any dissentient, let my opponent name him; for I have never met with such an opinion, or heard of its existence. Indeed there was very good reason, why the very notion would never occur to a theologian, of philosophical power (as such) augmenting our future beatitude. Observe, it is distinctly philosophical *cultivation*, as distinct from "the bare faculty" (p. 469), of which my opponent speaks throughout. But such cultivation, as I have already remarked (p. 18), is given by a countless series of actus intellectuales, each of which has its own position (better or worse) in the moral order. Many of these unquestionably have not been supernatural: and with the great multitude of philosophers, a very large number of such acts have been actually sinful; as directed to the ends of pride, vain-glory, inordinate ambition, and the like. To say that natural acts directly augment our beatitude, is open and undisguised Pelagianism; to say that sinful acts directly augment it, is an impiety of which no Christian ever dreamed.

Lastly, let me for argument's sake make the two extravagant suppositions, that Mastrius's opinion may lawfully be held, and that it leads legitimately to some conclusion at variance with my thesis. Still it would not at all follow, that Mastrian theologians can reasonably be cited as opposed to that thesis. No one even alleges, that they themselves *drew* any adverse conclusion, or that they thought of any contradiction between their thesis and mine. To know on which side their authority should be counted, is to know which of the two contradictory doctrines they would have retained, on discovering their mutual contradiction. If we are to exercise divination on this very uncertain matter, it is surely more probable that they would be ranged against my opponent; that they would adhere to that opinion, which they hold in common with all other theologians and which all regard as certain, than to that opinion which they know to be doubtful and to be regarded by many as unsound.

VIII.

I consider my third thesis then, as most abundantly and unanswerably established. And as I have already treated the fourth, I have fully vindicated those which, according to my original statement, "as they are prior in scientific order, so also are of far superior importance, to the remaining three" (Preface, p. vii.).

"The fifth question will be," I proceeded to say, "whether the doctrine contained in" my two first theses "be rightly and catholicly expressed as I have expressed it; viz. by saying that man's perfection consists exclusively in the perfection of his moral and spiritual nature," philosophical " excellence having no part in it whatever."

It will be observed, that I had myself stated this is a purely verbal question: and I may add, that if my four first theses be admitted, it is a matter of complete indifference to me, whether this fifth be accepted or no. I could almost wish indeed, that I were able to concede myself mistaken in this thesis, to save further trouble. But on the contrary, it seems to me that the very weakest part of my opponent's whole article, is that which concerns the present question; and I cannot dispense myself therefore from the duty, of vindicating my expression as true and Catholic.

But first I must earnestly complain, that he represents this purely verbal question as my "main position" (p. 468). Surely I have a right to be considered as the true interpreter of my own meaning; and in my Preface I expressly stated, that I regard this as purely a question of "true and Catholic expression." I am not seeking in this way to disavow a most serious responsibility for its use; because it is a great duty, not merely to express Catholic truth, but also to express it in a Catholic way. Still, at last, a question of words is one thing, and a question of realities another. And my opponent, by joining his chief issue on this question of *words*, has evaded the necessity of fairly confronting those allegations of mine on the question of *things*, which are in truth vitally momentous.

The first charge brought against me, under this verbal question, is truly remarkable. I was speaking, throughout my Essays, of such perfection as is attainable in viâ. This is abundantly manifest to any one who reads them, and will not (I suppose) be denied.

Thus, in my first Essay, I was contrasting the Catholic and anti-Catholic schools of thought. But no one ever supposed that Mr. Kingsley *e. g.* regards muscular power as part of our *future* perfection; and, as to several anti-Catholics whom I mentioned, there is no reason for thinking, that they have any belief at all in a future state. This, then, being the plain and undeniable fact, my opponent (p. 475) speaks of me as ignorant and unreasonable, because I use the word ' perfection' to express what theologians call 'perfectio,' instead of using it to express what they call ' beatitudo.' Here are his very words: " If we wish to hear what theologians have to say on man's perfection, we must turn to the treatise de Beatitudine." Observe, I expressly stated in my Preface, that this is solely a question of " true and Catholic expression:" and he maintains therefore, that my " expression" would have been more " true and Catholic," had I used the word ' perfection,' not as English for ' perfectio,' but for ' beatitudo.' It is quite notorious that, in theological language, ' viri perfecti' does *not* mean ' viri beati;' nor does 'status perfectionis' mean 'status beatitudinis.' And I should have expressed myself, not catholicly but *un*catholicly, had I confounded the two.

I pass on to other of his comments, which are at least on the surface less absurd than this. And at starting I make the obvious remark, that this phrase ' personal perfection' may be used in three different senses. (1) It may be used, to express the perfection of any power which personally belongs to me. In this sense, muscular strength is a personal perfection: so also is the power of writing, or dancing, or singing, or whistling well. (2) The phrase may be used, to express the perfection of any power intrinsic to my soul. In this sense, the power of readily reading French, or cleverly inventing or guessing riddles, is a personal perfection. (3) But there is a third use of the term, according to which "every creature is more personally perfect, in proportion as it more nearly reaches its proper end" (p. 49). I pointed out, that this is the sense in which theologians use the word; and as the whole question is one of "Catholic expression," there could not be a more complete vindication, for my own using it in the very same sense. Now the whole force (or rather weakness) of my opponent's criticism lies in his assuming, that what I say of personal perfection in this latter sense, is meant by me to hold of the two others also. Thus I stated, that our Blessed Lord "has, and can have, no Personal Perfection, excepting His Divine Perfections" (p. 86): and this answer fills my opponent with "simple amazement" (p. 486). I have more right to be amazed at his amazement. I would ask him to explain, what can be meant by our Blessed Lord, Who is God the Son, "*more nearly reaching His proper end*"? The perfections of His Sacred Humanity are His Personal Perfections, in the two first senses of the word: who ever doubted it? But then my remark avowedly referred to the third sense. And the third sense is not applicable, except to a creature.

Now I have already established my two first theses, which relate to the question of things: and I will beg my reader once more to read them. If then, according to the Catholic sense of this phrase 'human perfection,' man is more perfect in proportion as he reaches his true end, then I have "rightly and catholicly expressed" those two theses, by the language which I have used: I have rightly said, that man's perfection consists exclusively in moral and spiritual perfection, philosophical excellence having no part in it whatever. Man has many personal perfections, according to the two first senses of the word; in the third sense he has but one, viz. moral and spiritual perfection. In the first two senses of the word, philosophical excellence is doubtless a perfection of man: so also is muscular power; and the power of dancing gracefully; and the power of inventing riddles ingeniously. But theologians (as we have seen) use the word in its third sense; and (so using it) are unanimous in declaring, that man's perfection is measured by the degree of his charity. I have given my grounds for this statement in p. 48 and in pp. 80-82; and (incredible as it must appear) my opponent has not ventured to question it. He has not attempted (I say) to deny, that this very expression of mine, against which he so vehemently protests, is the one recognised theological expression. He says indeed that my view of things is (p. 493) "opposed to reason; to the very notion of man; to the language of theologians; to the Catholic doctrine" on the Sacred Humanity. And yet he has not ventured to deny either of the two following propositions. (1) In using the word 'perfection' to signify such perfection as is attainable in viâ, I have but followed the example of all theologians without exception; who reserve the word 'beatitude' to signify man's absolute and complete perfection. (2) In saying further that perfection (of this life) is measured simply by our degree of charity, I have again followed the example of all theologians, who are unanimous in the statement.

And here the force of my argument from our Blessed Lady is readily seen. Let the reader peruse carefully my former pamphlet, p. 86, and my critic's reply in p. 487. A knowledge of religious truth was of course part of her moral and spiritual perfection; and therefore (on my own principle) part of her personal perfection also. The question is, whether those perfections of the intellectus, which were *not* part of her spiritual perfection, appertain to her personal perfection. St. Antoninus and Billuart undoubtedly thought otherwise; and so I said.

My opponent argues, that on my view of the case our Blessed Lord is not a "Model and Example" to Christians (p. 486). Now, I will show in the next section that, from *my opponent's* principles, this really does follow, so far as the present life is concerned: here I will only say, that it certainly does not follow from mine. Christ is not, indeed, our Example and Model in this sense (nor would my opponent maintain it), that we in this life ought to seek the Beatific Vision; or to aim at "knowing all things in every way in which they can be known." But He is our Model and Example in this, that we (following most imperfectly and falteringly in His footsteps) should strictly subordinate all our philosophical exercises to spiritual regulation and control. His soul was unintermittently gazing on God; and His acts of contemplation were accompanied by equally unintermittent acts of charity. So far as He vouchsafed to elicit processes of reasoning, every such process was at every step commanded by charity, and directed to the greater glory of God. Oh, that all of us who philosophise, Catholic or non-Catholic, kept this great Example ever before our eyes!

And now lastly, as to that which my opponent seems to think his strongest ground of objection; the Catholic doctrine on Beatitude. For myself, I follow those theologians, far the most numerous, who consider Beatitude to consist formally in the Beatific Vision, and in that alone. But I am quite at a loss to understand his drift, when he adduces this doctrine in behalf of his own conclusion. He surely cannot mean, that philosophical activity is that process in this life, which most nearly approaches to the Beatific Vision hereafter. At least, if he do mean this, he runs counter no less to common sense, than to the unanimous judgment of theologians. St. Thomas, I suppose, may be accepted as a most fair representative of those, who regard future Beatitude as consisting formally in operation of the intellectus. But he teaches most expressly, that it is the contemplative life on earth, the prolonged and loving contemplation of God, which is the commencement and foretaste of Heavenly Beatitude.*

It may be said however, and truly, that I have not yet directly faced those emphatic passages on the knowledge possessed by Christ, quoted by my opponent from Billuart and St. Thomas (p. 485). The doctrine of those passages is the doctrine (I admit) of all theologians. It appertains, they all say, to the perfection of the human nature which God the Son assumed, that in that nature He shall know all existing truth in every possible way. The comprehension of God is indeed excepted; but (I think) nothing else. Now in strictness it is not *necessary* that I should consider these passages. For there is nothing in them which, even on the surface, appears contradictory to my first two theses; and my opponent himself admits, that my language about man's perfection is the one recog-

---

* "'Contemplatio Dei promittitur nobis, ut actionum omnium finis, atque æterna perfectio gaudiorum' (Aug.): quæ quidem in futurâ vitâ erit perfecta, quando videbimus Deum facie ad faciem; unde et perfectè beatos faciet: nunc autem contemplatio divinæ veritatis competit nobis imperfectè; . . . unde per eam fit nobis quædam *inchoatio beatitudinis*, quæ *hic incipit ut in futuro continuetur*."—2, 2, q. 180, a. 4, 0.

The whole of this 'quæstio' concerns the 'vita contemplativa;' and in my former pamphlet (pp. 84, 85) I quoted a passage from a previous article of the same question. In this St. Thomas quotes, with agreement, the statement of St. Gregory the Great, that such life *consists in* Divine Charity: "in quantùm scilicet aliquis, ex dilectione Dei, *inardescit ad Ejus pulchritudinem conspiciendam*." According to St. Thomas therefore, that life, which is the commencement of Beatitude, *consists* in Divine Charity.

nised theological way of *expressing* those two theses. Our present question, be it remembered, regards nothing but "true and Catholic expression." The only argument, which could imaginably be drawn against me from such passages as the above, would be the following: 'it is not probable, that theologians can speak of man's earthly perfection consisting simply in charity; since they say so much about knowledge, as being a perfection of the Sacred Humanity.' But my opponent himself virtually admits, that such an argument would be fallacious; since he admits, that theologians do speak of man's earthly perfection as consisting simply in charity. It is his business therefore, no less than mine, to harmonise the two statements; for theologians are equally unanimous in both. And as these passages cannot hurt me, so neither can they by possibility give any strength to my opponent's theory. For (as I will show clearly in the next section) that theory is destitute of all internal cohesion and consistency; his statements, when brought together, being manifestly preposterous and absurd. At the same time, no doubt these passages raise a question, which is of some importance in itself, and is at least closely connected with our present inquiry. I will therefore proceed to consider them.

The theological data, on which we have to proceed, seem to me mainly these four. (1) It appertains to the perfection of Christ's Sacred Humanity, that it shall know (with the above-named exception) all existing truth in every possible way. (2) Human perfection in viâ is exclusively proportioned to our degree of charity. This my opponent himself admits to be the universal statement of theologians. (3) In Mary there was no shadow of imperfection: yet theologians do not generally (I am not aware that any do) ascribe to her, while in viâ, any universal knowledge of truth; but exclusively, or almost exclusively, of such truth as was intimately bound up with her moral and spiritual perfection. (4) The evil spirits possess in some sense a very wide knowledge of truth: *e.g.* of psychological and physical truth. They know a great deal on the human soul; and a great deal on physical facts. Under the last head indeed, their knowledge (such as it is) is undoubtedly far more extensive, than any which is usually ascribed to Mary while a viatrix. And yet it would be the language of every good Catholic (unless my opponent be an exception) that they possess no quality, which is in the least admirable; that they are as simply grovelling and despicable, as they are odious and malignant. Now the difficulty is, how the first of these statements is to be reconciled with the remaining three; nor am I aware of any theologian who has directly treated it. In making therefore my own humble attempt, to work these four data into a whole which shall be consistent with itself and with reason, I put forward my remarks with the greatest diffidence, and with every submission to better judgments. Moreover, as the question is plainly a difficult one and I must be very brief, I will entreat the reader's active and careful attention.

We are more perfect in viâ, say theologians,—*i.e.* we more nearly approach the end of our creation,—in precise proportion as we have more charity. Yet we have numberless other perfections, besides charity, if we use this word 'perfection' in a different sense: for we may possess the power of writing, of singing, of wrestling well; and these are all perfections. For clearness' sake, I will restrict the word 'personal perfection' to the former sense; so that in viâ we have no personal perfection, except charity and all which charity implies. But what as

to those other perfections, which are not included in our
'personal perfection'? No one will say indeed that the
soul is in any sense more admirable, because we can sing,
or dance, or invent riddles well: these are rather perfections of some particular power, than of the soul itself.
But a real *knowledge of truth*, even though it gives us no
help towards spirituality, is a real perfection of the soul;
and the soul is more admirable for possessing it. It is
possible indeed (see p. 18) that the various actus intellectuales, which have accumulated such knowledge, have
been morally evil: and in that case, no one can call it a
perfection of the soul. Again, the very smallest increase
of charity is a greater perfection of our soul, than any
imaginable amount of knowledge merely as such: for it
must ever be the main excellence of a creature, quite
incommensurable with all other excellences, to approach
more nearly the end of his creation. Still, cæteris om-
ninò paribus, a real knowledge of truth is in itself an
additional perfection of the soul.

But can evil spirits possess a real knowledge of truth?
I think not. The obduration of their will causes the
deepest excæcation of their intellectus; and any true
apprehension of the Creator is impossible. This will be
admitted by all. But there can be no true knowledge of
the creature, without an apprehension of the Creator,
which shall be at least partially true; which shall be (to
use the common expression) true *as far as it goes*. For
the one essential truth about any creature, is its relation
to the Creator; those therefore who can have no true
apprehension of Him, can have no real or essential knowledge, in regard to the works of His Hand. The knowledge, possessed by evil spirits, is indefinitely available
indeed to them for practical purposes; but it is no real
knowledge of things as they are, and therefore no perfection.

If this theory be admitted, it will at once harmonise our four data. Such knowledge as the evil spirits possess, is no perfection, and nothing therefore which deserves admiration. Further, since our personal perfection is the 'unum necessarium;'—since the smallest increase of personal perfection outweighs in value the perfection of all imaginable knowledge, merely as such;—there was no reason of congruity, for God endowing Mary's intellectus with any knowledge, save that which contributed to her constant growth in personal perfection. But in the case of our Lord, there is no question of *growth* in love for God: His soul, from the very moment of its creation, elicited the very highest acts of love, which could congruously be ordained; and the intensity of His human love towards God knew neither increase nor diminution. His Vision of God, and intense love for Him, are placed by every theologian at the very summit of His human perfections, with which no others can compare: but it was also congruous, that He should possess every endowment, which is a true perfection of the soul.

What I have said, may be very easily applied to the question of philosophical and scientific power. Such power is not *in itself* a perfection of the soul; but only so far as it contributes to the *knowledge of things as they really are:* and in the evil spirits therefore it is no perfection; though they undoubtedly possess it in a state of very considerable excellence. Further, in none of us is it a perfection (but very much the reverse) so far as it has been generated by actus intellectuales which are morally evil. Thirdly, in regard to any one of us on earth, the smallest increase of love is a greater perfection, than any imaginable amount of philosophical power, merely as such. Finally, our Blessed Lord unquestionably possesses it in the highest degree; and it is a true perfection of His soul.

Such is the method which to me seems attainable, of harmonising the four theological data with which we started. But let me again pointedly remind the reader, that I have been treating the question 'ex abundanti;' and that my own various theses are in no way dependent, for their proof, on any theory which may be adopted. My remarks will perhaps have served their best purpose, if they attract the attention of some one more competent to treat the question, and if they thus prepare the way for a profounder solution.

IX.

And now let me contrast my opponent's view of things, as a whole, with that which I advocated in my former pamphlet. Surely so strange a medley was never brought together, as will be obtained, by combining the various propositions to which he has committed himself.

He earnestly contends (p. 472) that perfection of the will is no kind of security for perfection of the intellectus; and thinking so, he fully admits nevertheless (p. 475) that, according to the judgment of all theologians, our perfection in this life is measured exclusively by the degree of our charity. He must hold therefore that, according to the universal judgment of theologians, earthly perfection is fully compatible with our intellectus being in a state of deplorable degradation; that a man may be even fit for canonisation, whose intellectus is at the lowest, and who has therefore neither wisdom nor prudence except in the faintest degree. And this being the undeniable result of his own propositions, he rates me soundly for my disrespectful treatment of the intellectus, in the view which I take of human life here in viâ; and for my too exclusive regard to excellence of the will.

Even this is but little, compared with what follows. It is obvious to ask, if excellence of intellectus be no part of our perfection in viâ, why should we give ourselves the trouble of labouring at those philosophical processes, by which, according to his view, such excellence is attained? He virtually replies: 'because *hereafter* excellence of intellectus will be important, as the instrument of increased beatitude.' Do any theologians then, I ask, even dream of saying, that *acquired* excellence of intellectus will be an instrument of increased beatitude? There is no pretence for imagining that they do; nor *could* any one consistently say so, except the most abandoned heretic. See pp. 29, 30. What possible reason then can he give, for the importance of philosophical discipline? except indeed that, of its being a most useful instrument, for perfecting *the will*, whether of ourselves or of others; of promoting our own sanctification and the salvation of our neighbour's soul. But this is the very ground, on which I myself recommend such discipline; and which he regards as so derogatory to the intellectus, that he denounces my view of things as 'aggravating,' 'intolerable,' and 'unsound' (pp. 468, 493).

And further, according to the view which he attributes to all theologians, our Blessed Lord cannot be (in his opinion) our Example and Model while we are in viâ. For it is his very statement (p. 486) that Christ cannot be our Example and Model, unless philosophical excellence be part of our perfection; while he himself allows it to be the dictum of theologians, that in viâ our perfection is measured simply by our charity. Such is the strange exhibition in which my opponent's propositions issue: an exhibition (I venture to think) which would not have been presented, had he taken the trouble of pondering my doctrine as a whole; of considering the mutual relation of its various parts, before committing himself to an unqualified attack.

For my own doctrine is at least intelligible and consistent. I maintain, that "perfection of will and perfection of intellectus will necessarily proceed pari passu together." I hold that man's personal perfection, whether here or hereafter, in fact comprises *both*; consisting in the contemplation and love of God. Our earthly perfection indeed consists formally in love, and our heavenly in contemplation; but love on earth presupposes contemplation, and the Beatific Vision in Heaven necessitates love. If we are wise, we shall labour more and more earnestly at the great task, of subordinating every other pursuit to this our one essential work; the acquiring perfection here, with the hope of enjoying that Beatitude, which is our consummate perfection, hereafter.

If my opponent considered this theory open to just exception, he was bound patiently to think out some contradictory theory, which should be equally consistent with itself; and then to compare the two on grounds of authority and of reason. But it does not show a becoming respect, either to the importance of the subject or to the common sense of Catholic readers, when he expresses a very confident *negative* opinion, without having any positive tenets of his own: unless indeed we can dignify with that name, those crude and unmeaning extravagances which I just now recited. And still more should he have made himself sure of his position, before venturing to take the particular ground of attack which he has adopted. For his very complaint is, that I have been too precise for his taste, in following the language universally used by theologians on "perfection." His very complaint is, that I have maintained their usage of the word, as signifying the perfection attainable in viâ; and that (in accordance with their testimony) I have regarded it as precisely measured by our degree of charity.

X.

My fifth thesis has now been established; and on my sixth no question is raised. The seventh question remains: viz. "Whether those who appertain to what I have called the anti-Catholic schools of thought, really hold such opinions as I have attributed to them in my first Essay" (Preface, p. vii.). And this question includes two distinct inquiries. First, have I rightly interpreted one particular passage of Sir W. Hamilton, which I cited for a special purpose? Secondly, have I rightly apprehended the drift of anti-Catholic writers in general?

As to the former of these inquiries, the passage is quoted at length from p. 94 to p. 96. I must ask the reader, before proceeding further, to study carefully (1) the passage itself; (2) my own comments which follow; and (3) my opponent's comment on those comments: p. 467, 8. He will then see that the whole question turns on the issue, whether Sir William includes, or does not include, under the name "speculative knowledge," the whole philosophy of "moral, political, and religious truths," so far as the natural order reaches. If he do not include them, I have injuriously misrepresented him, and am bound to retract and apologise; if he do include them, my opponent has not a word to say in his defence. And yet, though this is so plainly the issue which my opponent himself chooses, I do not understand to this moment how the matter can be even doubtful. "Speculative knowledge," at the beginning of the first paragraph, evidently corresponds with "speculative truth" at the beginning of the last. "Speculative truth" includes the "mental philosophy" of a few lines later. And of this "mental philosophy" Sir William says, totidem verbis, that it "comprehends all the sublimest objects of our theoretical and moral interest; that every natural conclusion, concerning God, the soul, the present worth and the future destiny of man, is exclusively deduced from" it. My opponent reproves me, for choosing a discreditable interpretation "if another be open" (p. 494). I wish he would explain what other *is* open. It is idle, until he does this, to argue from other passages, that Sir W. Hamilton *could* not mean this: I ask the simple question, what other sense his words can possibly bear? Otherwise I should have much to say in behalf of my opinion, that the extracts, brought forward by my opponent in p. 490, are not at all unlikely to have proceeded from the same man, who wrote the long passage which I quoted in the sense which I ascribed to it.

As to my charges against anti-Catholic thinkers in general, in my Preface (which my opponent persistently ignores) I stated definitely what those charges were; I admitted my inadequate acquaintance with those thinkers; and I expressed my readiness to withdraw the charges on better information (p. xi.). My first charge was, that "all able and original anti-Catholic thinkers would expressly deny the Catholic proposition, that" philosophical "excellence forms no part whatever of man's true perfection." But as my opponent himself unhappily denies that proposition, he will not think me unjust to anti-Catholic thinkers, in saying that they also deny it: though I certainly think it a very heavy charge to bring against any Theist. "Secondly," I added, "the immense majority of them hold the implicit belief, that great" philosophical "power is in itself worthy of homage and reverence, quite apart from all question of the use to which it is put;" and from all question, I may here add, of the means by which it has been obtained. My opponent is strangely callous to the extreme odiousness of this notion (p. 466). In illustrating that odiousness, I will not attempt any personal criticism on Newton and Laplace, of whom indeed I know extremely little; but I will make a general remark. Nothing is more easily imaginable, than that the whole series of actus intellectuales, which have generated some astronomer's scientific power and knowledge, may have been sinful. I will not here speak of mortal sin; because to state and defend my convictions on this head, would lead me into a totally different subject: but at all events venially sinful. Nothing, I say, is more easily imaginable, than that all those actus intellectuales have been motived by pride, or vain-glory, or ambition, or sinfully inordinate love of knowledge, or the like. Turn now to Lord Brougham's rhapsody, quoted in my former pamphlet, p. 20. No one will doubt, that he is speaking simply of the great scientific genius displayed by Newton and Laplace; and that he would say the very same thing of our imaginary astronomer, were such astronomer's genius equal to theirs. If we start then from Lord Brougham's principle, it is abundantly possible, that one protracted course of venial sin may "almost exalt the nature of man *above* its destined sphere," and raise the sinner to "a station apart, *rising over all the great teachers of mankind*, and" with justice "*spoken of reverently as if*" he were not a "*mortal man*." My opponent admits, that Lord Brougham is here "*somewhat* extravagant;" he is resolved to be on the safe side. Protestants charge us, I added, with idolatry towards our Blessed Lady; but at all events the object of our veneration is one, whom we believe to be eminently holy. It has been reserved for enlightened Protestants, who have indignantly repudiated 'creature worship,' to bow "reverently" before that, which is compatible with any given amount of sinfulness and depravity. Far be such "foul and degrading idolatry" from all good and loyal sons of the Church!

My statement in the Preface was, that, according to

my present impression, the "immense majority" of anti-Catholic thinkers would "habitually speak of" philosophical excellence "in that reverential tone, which is not exemplified indeed, but caricatured, by Lord Brougham's wild extravagance" (p. x.). I added, that my whole statement referred, not to their explicit language, but to their implicit tone and drift. My opponent, in defending these thinkers, was called upon to reply 'Yes or No,' to this my impression. Does he really mean, that Sir William Hamilton, or Kant, or Sir L. Bulwer, or even that Dr. Whewell, is free from this reverence for philosophical power as such? It is not worth while to examine the particular citations which he adduces, until I can gather from him, whether he even professes to deny the general charge which I bring.

He quotes with approval, *e. g.*, a truly admirable passage from Kant, endorsed by Sir W. Hamilton (pp. 490, 491), enforcing the truth that my "moral worth" should be "the absolute end of my activity." Does my opponent even allege, that the general drift and tone of Kant and Hamilton is in accordance with this admirable maxim? That they consistently regard all human "activity" as misguided and valueless, which is not directed to "moral worth" as to its end?

Indeed, as far as the mere statement is concerned, I cannot see where Kant's maxim differs from the Catholic doctrine, that "my earthly perfection consists exclusively in the perfection of my moral and spiritual nature." Take the particular case of philosophical cultivation: what is the difference, between the "directing" such cultivation to "moral worth" as its "end" on the one hand, and the making it an "instrument of moral and spiritual perfection" on the other? Why does my opponent regard that doctrine as admirable in Kant and Hamilton, which is so "unsound" and "untenable" when stated by a Catholic? I believe the only real difference between the two cases to be, that the Protestant only says it, while the Catholic means it.

As to my Essays being "singularly ill-timed" (p. 465), no one surely will say that they are ill-timed, if their doctrine be sound. It is an elementary truth—the starting-point of the Catechism, the foundation of all spiritual exercises,—that man's end in this life is the loving and serving God; and that we simply err from our true path, if we prosecute either philosophical studies or any other processes, except as instruments to that end. My opponent's criticism in itself shows, that this elementary truth is forgotten,—nay, is denied,—by some of those whom it most concerns. I have surely performed an important service, if I have drawn their attention to this truth, and succeeded in vindicating it from misconception.

Probably enough it is desirable, that this question shall be still further debated; and that various principles, which have been implied on either side, shall be explicitly stated and discussed. But if my opponent returns to it, I entreat him to take my original theses one by one, as I myself stated them. Let him explain, both to himself and to me, the precise ground which he occupies, in regard to the first; to the second; and so on with the rest. Especially let him fairly meet (instead of totally ignoring) that long passage from p. 39 to p. 47; which I expressly represented as the basis of my whole argument. In no other way is it possible, for him and me to arrive at a mutual understanding.

www.ingramcontent.com/pod-product-compliance
Lightning Source LLC
Chambersburg PA
CBHW031551110426
42739CB00039B/1075